HOMESCHOOLING MADE EASY

HOMESCHOOLING MADE EASY

AVERY NIGHTINGALE

CONTENTS

CHAPTER 1

Introduction

The educational process should be an invigorating one. After all, preparing children for a life of successful independence is an important part of the reason you took on the satisfaction of educating your children. But while you can see that teaching children good character by monitoring their habits and attitudes is all-important, take heart. Choosing curriculum materials from the many available publishers is your launch into the process of equipping your children with a lifetime love of learning, the specific subject area skills they may need for independent study, a healthy foundation physically, and a storehouse of productive habits.

Homeschooling can be challenging. Yet, despite this challenge and the many expenses and time involved, a growing number of parents are taking on this responsibility because they value the education and unique social opportunities they can provide their children. If you are a parent, creating a stimulating atmosphere at home will maximize your efforts. Focusing on your child's innate curiosity and providing a variety of exciting activities linked to concepts can help you ensure educational success for your child. We hope that this guide will give you a useful template for the joyous activity of teaching the children who depend on you.

1.1. Purpose of the Book

Homeschooling your children can be very satisfying and surprisingly effective when parents are aware of the needs and achievements of their children and are motivated to contribute to their own families' goals. The family's goals are likely to change over a period of time, perhaps even several times per year. During such transitions, you might find more opportunities for family closeness or for meeting the academic goals of one or more members. Would you like to explore, and then actively pursue, the best learning possibilities for yourself and your children? There are many free, low-cost and high-cost documents, programs, and curricular paths available to you. You can learn how to recognize and use these numerous options for meeting your own family's unique needs by reading this book.

The purpose of this book is to inspire and coach parent educators to be their best. Do you make yourself feel guilty by comparing your mundane days with some neighbor's glamorized success? All over the nation, families are choosing to educate themselves for reasons as diverse as life. From overcoming serious personal tragedy, isolation, or social pressures to more traditional motives of providing the benefits of personalized education, enhancing closeness with family members, or addressing a specific academic or socio-academic concern, homeschooling families share many of the same lifestyle goals as their next door neighbors and yet develop these goals through a very different route.

1.2. Benefits of Homeschooling

Florida homeschool regulations require that the child be taught regularly for at least 180 days, Monday through Friday, in specific courses in the following areas: language arts, mathematics, social studies, science, United States Constitution, health, reading, spelling, residence, and self-assessment. During this year, 60 days per

180 days may include the hours of attendance during which students participate in religious instruction as part of a home education program.

Parent educators often know and understand their children better than anyone else. Knowing the interests and learning needs of your child, parent-educators can motivate him and adapt lessons with his unique state. Our children are not subject to learning disabilities, poor teaching methods, teaching aids, or a system that doesn't meet their needs. They are allowed as long as they need to learn the material without feeling useless or that they are keeping others waiting.

While some of us worry that our children may have missed out on some things that their friends do or learned in other schools, experience teaches numerous benefits of home education. Some of these benefits include having the freedom to choose from a variety of books, learning materials, and educational methods rather than thinking that some big textbook company should dictate your choices. Children learn more valuable lessons in a short time learning one-on-one than in a full week institutionalized classroom. Both the parent and child can use their analyses or questions in their daily lives as teaching moments. Parents can see their son when he "gets it" and then move to the next area instead of wasting a large portion of the day teaching lessons he has already mastered. With homeschooling, you do not even begin teaching the Bible. You have the opportunity to see the "light" and one day understand their lessons.

CHAPTER 2

Getting Started

Homeschooling is not right for all families. There is no question that there are drawbacks to home education. At the same time, there is no question that there are shortcomings to public education. Every family is different. Any decision about how to share educational values and truth with your child should be made after considering the pros and cons of home education. Let's take some time to consider the top reasons to homeschool. Your list may be different, but thinking about your goals and needs can help you get started on determining if preparing and teaching your child at home will work for you.

Many families find homeschooling the best choice for their family's educational needs and the family's lifestyle. This choice is unique because parents make the decisions that shape their children's education. Most parents also decide to homeschool for more than just academic reasons. In fact, many parents want to develop a close relationship with their children. Parents and children can form lasting, meaningful relationships. Families can work together, travel together, and share a common purpose. Parents have the opportunity to help their children develop morals and values. These social benefits create a strong family unit where everyone involved has a sense of belonging and shared values. When your family decides to homeschool, homeschooling is another star in your family constellation of activities and interests.

2.1. Legal Requirements and Regulations

Parents who intend to teach their children at home are required to notify the Arkansas Division of Elementary and Secondary Education. Parents must provide proof of a high school diploma, and all conventional certified teachers wanting to home educate a student must be certified. Parents must submit a curriculum with the application to the district school superintendent for approval. An annual letter of assurance must be submitted to the local superintendent stating any changes, enrollment withdrawals, or completion of the annual assessment, including the quarter in which it was taken. Parents must ensure that the child is in compliance with the state-required immunizations, health screening, and submit a criminal history of any individual who works directly with the child. Evaluate the progress of the child academically. Children in homeschools are allowed to participate in physical education, art, music, or drama classes and activities offered by the local school district. This does not mean they or other homeschoolers are part of the student body. Religious classes are excluded.

Homeschooling is legal throughout the United States, but each state has different regulations concerning homeschooling. It is very important to understand your state's law and the local school district's requirements. Find out what is required of homeschoolers and follow the law. Generally, states need to see that your child is being instructed in subjects commonly taught in the public schools of the state and other subjects related to your child's needs. Plus, you are paying more than lip service to education. States differ on the need for tests, home visits, portfolios, or submitting objectives, and some other details. While finding out the regulations, remember first that homeschooling is legal and second that homeschooling is not a division of a public school (unless your state specifically empowers the local school authority to supervise you). Be careful to use the worldview that homeschooling freedoms are civil liberties, since homeschooling laws are based on principles of freedom and not on a needy dependent relationship between the homeschool and the local public school.

2.2. Choosing a Curriculum

There are many learning goals around which we might want to build a curriculum (family religion, family ethics, peer relationships, hobbies, sports, character development, etc.). When we speak of curriculum in this article though, we are thinking about materials for education in academic subjects. At times, within a given subject, we may want to focus on certain goals or skills. For example, we might want to develop a domestic science experiment plan as part of a chemistry course. These would, of course, look quite different for the teen who wants to enter a science fair and the teen interested in the culture of life. There would be a natural desire to cover some similar general areas in these two education plan examples. Educational goals, learning methods, types of resources to be used, methods of assessment (testing), etc., will not be the same from plan to plan. The plan developed by each educator should be unique to himself or herself and his or her student.

The importance of choosing good curriculum materials cannot be overstated, especially for new homeschoolers. A curriculum is a plan of study. It spells out what resources for learning we are going to use to achieve learning goals. We need to have some idea of what these learning goals are and how to match our curriculum products, as closely as possible, to the many unique, individual needs of our own children. We can't do it all, so we must sometimes choose between many good options. It would be hard to imagine a professional educator not using a curriculum. Choosing good curriculum materials means we have some faith in what we are using - that the printed materials, software, internet resources, etc. are going to be an efficient and effective means of helping educate our children.

Setting Up Your Homeschool Space

Most young students need help from parents to complete their homework. Ideally, they would have a designated place to do this work every day that is set apart from the rest of the family activity. If your family does not already have a place like this, consider setting up a space specific for your child's homework near where you are working. This will allow you to assist your children while still participating in the whole family dynamic. You may also think about creating a homework system that fits your family. Scheduling a family homework time with your children may be beneficial. If you start the space arrangement off right, it will continue to function for many years to come. So take the necessary steps to establish a space that promotes a conducive learning environment.

Deciding how your space will function is important. Have one designated place where school materials are kept so that you are not constantly searching for lost items. It is helpful if this space is away from the living/family room. In addition to a designated school area, choose a spot for morning basket time that is in the main area of the home; have a designated place for coats, hats, shoes, school bags, etc.; and make sure each child has a space of their own (like a little desk, a school storage crate, etc.) while working on their various projects and homework. It is important for children to have a place where they feel comfortable while working and organizing, a space that makes them feel that their studies and work are important. Make an organized place for parents to do schoolwork too. Buy file crates to keep children's completed work, projects, and reports in.

CHAPTER 4

Creating a Schedule

You can either start with a long list of what to do when, or a really short list. I start with a really short list because I like to keep things simple. Then, once I get that list down, I can add as necessary and also stay away from coloring everything with so many rules, records, and charts that chaos is inevitable. A weekly schedule is a great practical name for what happens in a week during your schooling and life. You can combine both into one schedule and see what time is left over for more. Or you can have two schedules, one for life things and one for schooling things. Each item on your schedule can be as detailed as you want - you can detail every meal, story, and lesson. Or you can just have loose sections for morning, afternoon, and evening and do your life randomly within those time frames. Save the hourly time tables for school children in private and public schools.

The choice is yours. It's okay to school for a shorter or longer time, take longer or shorter breaks or mix up the day. Just do what works best for you and your family. Remember, homeschooling is as individualized as the families doing it, so be open-minded and flexible. Traditionally, homeschooled students have studied and worked during the day while their school friends are in school. This leaves our afternoons free for visits to the park, creative activities, field trips, or other fun and learning experiences. This goes back to the idea of homeschooling being focused and individual. It's one of the great opportunities available to homeschooling families.

CHAPTER 5

Effective Teaching Strategies

Thirdly, give students some feedback. As a pedagogical procedure, students were assigned examinations for assessing their learning ability. Whatever effort is done by the students is recorded, tally the highest exam, and designate grades based on the scores achieved. Assignment work scores are used for measuring the understanding level of the students.

Secondly, present information frequently to the students. How do you know that I am learning computer skills? You know that you are learning the computer when you are engaged in the practice of keyboard works in terms of knowing introductions to computers via rich media, diagrams, display presentations, etc. These forms of visual presentation will certainly enhance the student's computer applications.

First, you need to think and organize skills and concepts before teaching. What concept should be taught? How will skills be organized? A pre-planned database of basic skills and concepts is known as a "syllabus". Hence, a proper syllabus and these predetermined targets can help the parent to achieve them according to the design and method of the system.

Effective teaching strategies to use with homeschooled children are essential. Ironically, teaching strategies are not rocket science; on the contrary, they are quite basic. We already know them to some extent because we have all been taught. As an educator, you need to keep the following points in mind.

CHAPTER 6

Incorporating Technology

A computer can also be a key asset in creating a student newsletter to share with family members. This newsletter can include what is being taught and what the student has learned. It can include a computer-assisted photo album complete with student captions and comments. In turn, family living at a distance can use the computer to send email. Also, someone in the family can provide live-time mentorship to their student from abroad with the help of the internet and simple videoconferencing technology. If automatic dial-up is unavailable and the phone line is free at night, an answering machine and the internet can be used to locate free midnight email services to send and receive messages. However, if either email or an internet message board is used, it is important to teach children good habits in their communications including to whom and why they are communicating, and how to avoid objectionable material.

A good use of technology saves homemaking parents time and energy. Besides being used to educate their children, technology can be used to simplify recordkeeping and communications. A simple database program can be used to keep grades, family and student records, and lesson plans. Information in the database can also be used for student evaluations and transcripts for high-school students. Also, recordkeeping of grades can be kept in gradebook software.

CHAPTER 7

Assessment and Evaluation

Select tests suitable to the material that has been studied. There are tests that accompany certain books, or you may be able to create your own. Use the review workbook already mentioned as a source of such tests. For students age 9 and up, consider using those available through the A.C.E. program which may be ordered by the subject quarter. These final tests, together with the materials, readings, speeches presented, and projects completed, are very useful for teacher and student preparation for annual evaluation. At the end of 3rd and 4th grade, and at 3-year intervals after that, all Colorado students, including those being homeschooled, must be evaluated by achievement tests. This evaluation is necessary to assure that students are benefiting from instruction, and, together with the annual progress report, is to be included in the homeschool records.

The topic of testing stirs much discussion among many homeschoolers. Few doubt that their children's work should be evaluated, but finding an effective way of assessing learning is a different matter. Why is it important? First, testing is the child's learning. Just as he needs to be reminded that he cannot go on to the next level of math until he can add a sequence of numbers that sum up to 9, so also must he be tested at least occasionally to see if he has comprehended what he has been taught. Second, testing is a good discipline for children, especially if it is given at regular intervals and fits well the material that can be studied. Third, testing is necessary to summarize the material learned in a subject and to help the teacher know if the child is progressing. A day of reckoning comes eventually, and it is better to know early on how things are going.

CHAPTER 8

Support Systems for Homeschooling Parents

One key to successful homeschooling is for the parents to provide the support the child needs. This also includes providing the support that the homeschooling parents need. Two areas are important. The first is to remember that homeschooling is different from traditional schooling and that many things can be tailored to the way the parents like. There are many flexible alternatives available. The second area is to realize that, because most parents are unfamiliar with homeschooling and the many excellent tools available today to assist parents, they tend to become overworked too easily. Even interested homeschooling parents can quickly become overwhelmed with a sense of responsibility. Under these types of conditions, homeschooling parents can degrade into not providing the proper academic, emotional, nor physical support to the children, and the process can rapidly spiral out of control.

Homeschoolers are, by definition, understated. Ask any homeschooling family what they do, and invariably if they describe it as school, they refer to their school schedule, homeschool lessons, etc. We have observed a strong tendency in homeschoolers to make extraordinary efforts to appear normal; to look like traditional school, despite the well-known deficiencies in the traditional education system. On the other hand, what makes homeschooling special is the freedom from the artificial timetables and the discredited techniques of traditional schooling. There is no need to proceed like traditional schools. There is no need to follow an 8 am to 3 pm schedule. There is no need to teach or study each subject an arbitrary number of minutes each day. There is no need to follow the traditional district curriculum.

CHAPTER 9

Dealing with Challenges and Burnout

Sometimes, formal physical and emotional care, organization, and from-the-community support do not prevent frequent exhaustion and the symptoms of burnout from setting in. The parent teacher with these symptoms would benefit from listening to other perspectives, that is, the children's wishes regarding the lesson and hearing their previous experiences. It would also be a good idea for the parent teacher to visit other more relaxed homeschoolers, take short and frequent breaks, and other means to prevent burnout. Before the undertaking becomes a callous and negative task, the child might have to go to regular school for a school break. If burnout has aggravated, however, it would be better to enroll the child sooner. There's a saying, "The best choice is the child's choice." If burnout becomes a problem, parents of the homeschooling child would benefit all around if they allow the child to have their say and have them back on track as soon as possible.

Staying energized and motivated is a challenge in this undertaking, especially for the primary parent teacher. Physically, home educators have a demanding schedule; they are responsible for the homemaking, nurturing, disciplining, teaching, driving to extracurricular activities, and continuing with their other works. The first line of defense should be good physical care - adequate rest, good nutrition, and regular exercise. Emotional support can come from being part of a community like the local homeschool association. Socioemotional issues are discussed and knowledge and materials are shared with frequent meetings and field trips. The homeschool association is also in a unique position of relating official information from the Department of Education. Realistic, flexible goals and planning and organizing the day's schedule also help.

CHAPTER 10

Special Considerations for Different Age Groups

Preschoolers - Preschool children love to learn. Learning is fun for young children and lessons that are fun will be more effective. Use simple lessons that will be most effective. Since they must master the basics, spend a lot of time with them to encourage their mastery of the 3 R's - reading, writing, and arithmetic. Patience and persistence are the necessary tools for your success. In preschool, focus on a few activities that produce great results versus a full day of schooling. Kindergarten should not be any more difficult or complex than preschool. Use plenty of encouragement and praise and don't demand perfection. Your patience and persistence will lead your student to success. For children who easily learn their tasks, use additional busy work time. Keep it simple and always test any additional curriculum for mastery of each student. These activities may be things such as developing a hobby-time, pursuing musical lessons, and adding additional extras to the schedule.

As parents, we know that each individual child is unique. This uniqueness is evident during the homeschool year. Part of a successful homeschool year is recognizing that each child has different needs. We also parent multiple ages of children at once. How do we accomplish this? What can children handle at different ages and how does it change?

CHAPTER 11

Homeschooling and Socialization

There are co-ops for shared teaching, classes for specific interests (Lego robotics, painting, public speaking), Latin and history and drama clubs, writing labs and math clubs. Large homeschooling resource organizations or networks are springing up in many areas. These organizations are vehicles for looking for social opportunities; offer support networks, classes in a wide variety of subjects, chorus and orchestra, drama productions and study halls; and have lists of families willing to share transportation. After-school extracurricular activities provide a great way for children to meet and work with others. Many homeschooling children, especially by the time they reach the upper levels of high school, regularly work together in classes which are organized by position in the mental pecking order, not by age. They may be the only sixteen year olds, but they're with other students who know as much as they do. Many post-high school students actually do begin taking college classes on a part-time basis while finishing high school.

Multiple research studies show that homeschooling produces well-balanced, functional human beings. This shouldn't be surprising once you consider that traditional schools have only been around for just a few hundred years while homeschooling persisted for thousands of years. Who's going to provide a better social training program: a twelve-year-old on the playground or a professional educator? And after all, homeschooling today provides an abundance of social opportunities for even the most wary parent. Volume One of The Well-Trained Mind details dozens of easy, low-stress ways to provide regular interaction with children of the same age, including ideas for play groups, book clubs, 4-H activities, scouting, team sports and other athletic groups, music ensembles, and science clubs.

CHAPTER 12

Balancing Homeschooling with Work and Other Respon

1) Plan comprehensive educational activities 2) Set family and individual goals 3) Coordinate learning activities 4) Make schoolwork portable 5) Get professional help if you need it.

Homeschooling enables children and their instructors to take advantage of a wide range of educational venues, such as hands-on museums, libraries, field trips, travel, and volunteer work. To succeed as a teacher, you must also learn to be a master juggler, balancing the needs of your students with everyday responsibilities, such as earning a living and taking care of home and family. With planning and practice, parent educators who work or have other pressing responsibilities can help their children get a good education and their own needs met too. Five of the 12 dozen tips you can use are:

Some parents who live in a house travel often or work at home, while others may be employed by an outside organization or business. While it is easier for a stay-at-home parent educator to teach his children, a working parent can homeschool too. The key to effectively teaching your own children despite a busy schedule is to balance your family's many responsibilities. Consider the following practical tips designed to help you blend homeschooling with everything else you and your family need to do.

CHAPTER 13

Resources for Homeschooling Parents

Aside from books, you can get planners and organizers that aid in record keeping. Workbooks and activity books provide important learning experiences, and the fact that once the books have been completed, they can be saved and referenced later makes them an awesome resource. Electronic resources are becoming better and cheaper, so many parents are considering homeschooling in part because of the technology available. Manufacturers are now marketing software that allows kids to interact with fun characters - a new factor in homeschool education that children brag about when they talk to their friends. For example, an instructional tape or interactive CD that accustoms kids to using a computer makes the remainder of your subject matter experience much easier. Although the right software can be very educational and enjoyable, parents should not use it as a babysitter. If used too frequently, it doesn't connect the children nor oversights their education.

The challenge of creating lessons by themselves and their responsibilities to guide the use of pencil and paper can overwhelm parents who are beginners at homeschool education. However, advancements in technology have now brought homeschooling support within easy reach of today's homeschooling families. Resources to help you can be classified into print, electronic resources, and organizations. Print resources include books, charts, educational aids, workbooks, and activity books, etc. that supplement learning. Virtually all educational publishers offer books that parents can make use of to develop a good homeschool program. Although you can no longer ask Dr. Spock, you can still ask your librarian. There are as many reference books out now about teaching as there were those years ago about child rearing.